A Living Organism Called Earth

The Geophysics of the Planet

Copyright © 2020 JOSÉ RUIZ WATZECK

English Version

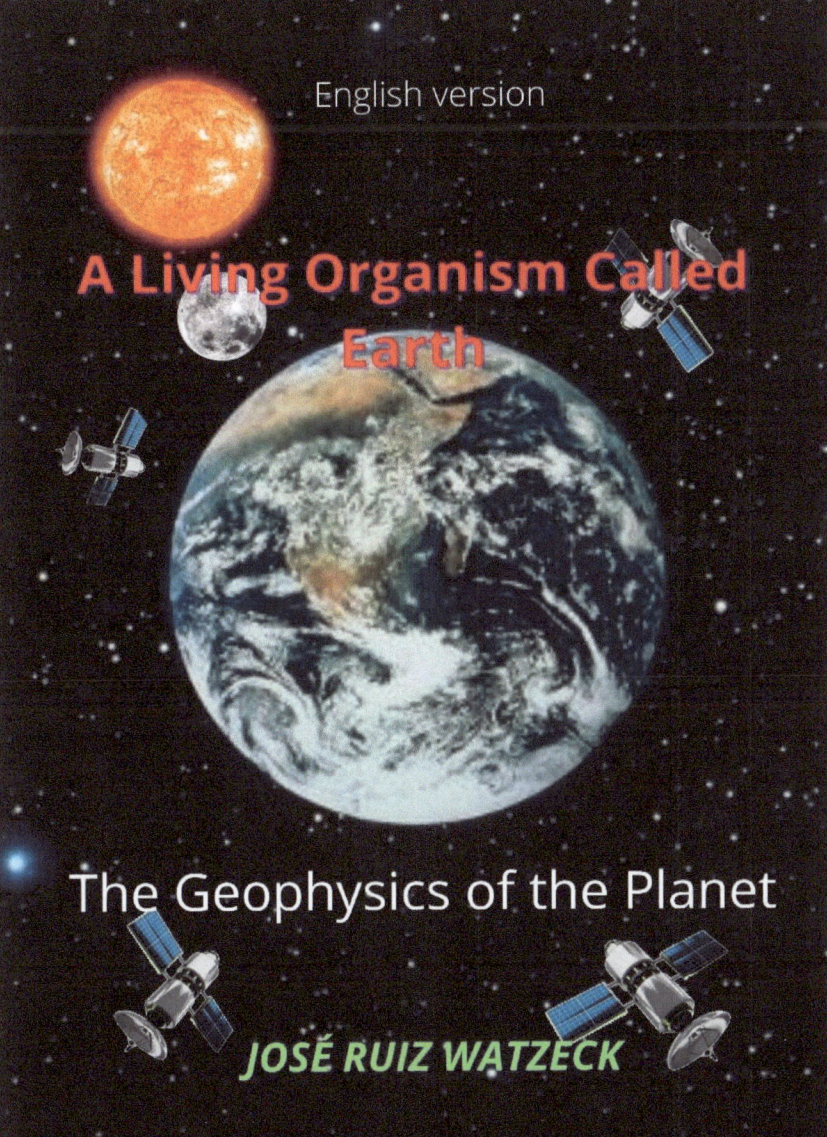

W353o   Watzeck, José Ruiz, 1977

        A Living Organism Called Earth - The Geophysics of the Planet - José Ruiz Watzeck

        1st Edition - São Paulo, Brazil 2020.
        67p. 22,86 cm
        English Version

1st Geopolitics. 2nd The strategic value of the ionosphere.

        I A Living Organism Called Earth - The Geophysics of the Planet

        CDD: 550

Summary

Chapter 1- The Storms 1

Chapter 2 - Antarctica 8

Chapter 3- Planktons and Phytoplanktons 21

Chapter 4- The Amazon Forest 29

Chapter 5- The Fire 33

Chapter 6 - The Sun 38

Chapter 7- The Earth Atmosphere 45

Chapter 8 - Human Beings 49

Bibliographical References 52

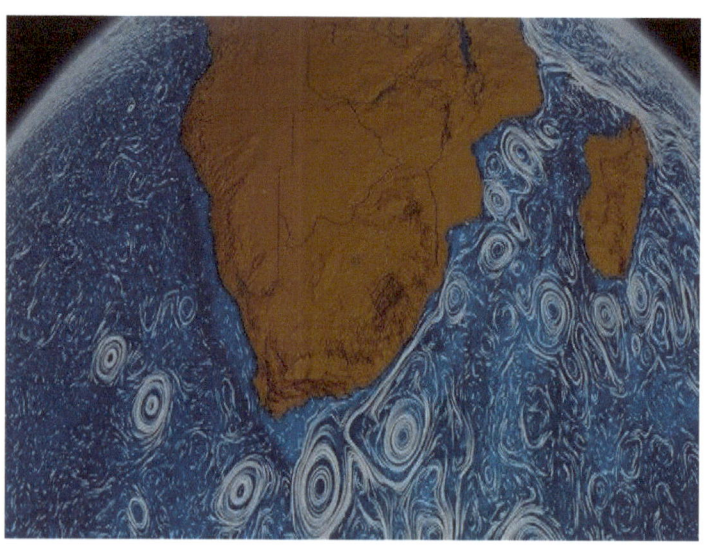

Foreword

Natural and hidden phenomena that devastate our planet, now, thanks to the most sophisticated technologies, allow us to study them in an unprecedented way, satellites scan the entire planet and reveal an enormous wealth of details. Never in the history of humanity have we had an account of this planet, a living and dynamic organism with highly relevant properties. In this work, we will know how the whole planet is interconnected, how everything is intimately connected, from one point to another of the globe, through technology, we will make a dive into the oceans and together, we will understand what the Saharan desert interferes with the Amazon, what the huge ice platforms in Antarctica contribute to maintaining a harmonious climate of ocean temperatures, because the fire naturally produced helps for the renewal of the most different types of life on Earth, how and why they occur at dawn, how the global cline really works, in which, the maritime currents interfere in the distribution of heat to the hemispheres. Let's understand why one of the Earth's layers known as the Ionosphere, formed of Hydrogen and Helium acts as an electric conductor, distributing all the lightning charge in the atmosphere of the entire planet. The chemical reactions of the clouds, and what the electrical discharges have to do with the formation of Nitrate. As these satellites show us the energy emitted by our star, the ultraviolet radiation, fractions of protons, electrons and neutrons discarded by space, electromagnetic pulses and the ejection of coronal mass. How the magnetosphere protects the Earth from strong solar blows.

From now on, we will count on the help of a set of satellites, so that we can in a scientific way, understand how our planet works. Every second, these equipment's record, measure and transmit thousands of terabytes of data, and only with this data, we can, for the first time, make a digital analysis of planet Earth.

In order for us to sequence this study, we need to know which are these tools that are orbiting the Earth, which if they did not exist, this study would never be possible.

The first satellite that helps us understand the climate is **Earth (EOS SER-2)**, a multinational NASA research project main focus is *Earth Observing System* (EOS). The satellite was launched at Vandenberg Air Base on December 18, 1999, on board the Atlas

II, and began collecting data on February 24, 2000 (EOS). The Earth carries a load of five remote sensors, designed to monitor Earth's environment and climate change. This satellite resulted in over 15 years of analysis and data collection.

The other satellites are **Aqua (EOS PM-1)**, a multinational survey of satellites in orbit around the Earth, designed by NASA, with the objective of analyzing precipitation, evaporation and the water cycle. It is the second main component of the Earth Observation System (EOS) right after the **Earth** (launched in 1999). Aqua was launched on May 4, 2002, from Vandenberg Air, on board a Boeing coupled to a Delta II. The helium-synchronous orbiting satellite. It orbits at an altitude of 705 km leading a formation called "train" with several others satellites *Aura, CALIPSO, CloudSat and the French PARASOL*). It has six instruments for studies of water on the surface and earth's atmosphere.

Aura (EOS CH-1) is a NASA multinational research project. The Satellite is in orbit around the planet Earth, analyzing the ozone layer, air quality and climate. It is the third main component of the

Earth Observing System (EOS), being first two:

EARTH (released in 1999) and Aqua (launched in 2002), respectively. The name *"Aura"* comes from the Latin word for "air". The satellite was launched a t Vandenberg Air on July 15, 2004 on board a Boeing Delta II 7920-10L rocket. The Aura orbits with the so-called "A- Train", a set of several other satellites carrying four instruments for studies of atmospheric chemistry.

We also have the **SDO (Solar Dynamics Observatory)**, an unmanned probe from *NASA,* which studies the Sun's processes that directly affect life on Earth, and whose launch took place at Cape Canaveral on February 11, 2010. Containing four telescopes embedded in its structure, two solar panels and two long- range antennas. Among its main instruments are the *Extreme Ultraviolet Variability Experiment,* which will measure the ultraviolet irradiation of the star in high definition, the *Heliosismatic and Magnetic Imager,* which will study the variation and characteristics of the solar interior, and the components of magnetic activity on its surface. In addition, it carries the revolutionary *Atmospheric Imaging Assembly*, capable of transmitting images of the entire solar disk, in strips of ultraviolet and infrared, not reached before by their predecessors.

Chapter 1- The Storms

In August 2005, about 400 kilometers off the northwest coast of Africa, in a volcanic archipelago, is located the island of Cape Verde, the hottest time of the year, in a period of every 72 hours, storms agitate the local ocean waters. A cluster of huge clouds begins to form, a vast event that will affect the entire world, only with the last word in space technology was it possible to understand such phenomena. About 700 kilometers high, the *Aqua* satellite registers an elevation in water temperature, with an infrared scanning system, points out that the ocean has reached the critical temperature of 26°C, with large areas more heated, begins to evaporate very quickly, this vapor absorbs the heat of the ocean transferring immediately to the air. With large capacity, water begins to carry energy, which will unleash in total destruction elsewhere on the globe. The specificity of this satellite (*Aqua*) to track water vapor shows us only a small specific scale of an interaction between ocean, air and sun, without any human being being being able to see the naked eye. About 200 tons of water are evaporated per hour. A process that consumes energy compared to a modest nuclear power plant, 1000 meters above, this vapor is condensed into cloud shapes, releasing heat and intensifying the air temperature by several degrees. As the air heats up, powerful vertical winds begin to produce, raising these clouds to approximately 15 kilometers in height, as the storm cell increases the effect of the Earth's rotation on the force to rotate. These gigantic clouds, merging in circular form, at this moment we witness the birth of a hurricane. With the data sent by the satellites, we can conclude that a hurricane is an immense power plant produced by nature. Being monitored and accompanied by ISS (International Space Station) and translated into Portuguese (International Space Station), the hurricane moves quickly across the Atlantic towards the Southeast North America, in a few hours it enters the Gulf of Mexico, where warmer waters enhance the storm. Right now, we can say that the people of this location are about to witness the power of the sun in the ocean.

At this moment, one of the most devastating hurricanes in the region, *Hurricane Katrina*, a tropical storm that reached **category three on the** *Saffir-Simpson* land scale and **category** *five* in the Atlantic Ocean, with gusts exceeding 280 kilometers per hour, with lower pressure of 902 mbar1, left the number of 1,883 people dead and reaching the areas of Bahamas, South Florida, New Orleans, Alabama, Mississippi, Louisiana. This is the physical capacity of water to retain and release energy. However, devastating this phenomenon has been to local people, the world owes its lives to the process that produced the storm, for the simple reason that when the ocean reaches a temperature too high, these storms are its escape valve, redistributing heat around the planet and balancing the global climate. This specific hurricane helped to cool extensive swathes of the Atlantic to over 4° C, rebalancing the ocean. And this phenomenon is just a small detail in an extremely complex and through the satellites we can affirm that everything is interconnected in a planetary way, literally, it is these hidden connections that keep us alive.

As the Earth circles its axis, several satellites record and analyze numerous data such as temperature, electrical charges, pressures and even the slow process of continental drift. Through technology, we can understand why parts of the plant are fertile and others completely dead.

São Paulo, month of June, 22° C, the citizens start another day of work, with winds below 12 km, a little more than 14.000 km of this point, in the city of Delhi in India, the inhabitants suffer with torrential rains, in few minutes, the streets become flooded and impassable, in this same instant, a forest fire devastated the North of Australia and on the coast of China more precisely in the city of Shanghai, hailstorms punish the region.

Before the technology, such events seemed to have no connection between them, when in fact, they are all interconnected. With the crossing of data from five different satellites, it reveals a layer of the system, the dynamic atmosphere that encapsulates the whole world. With all this data, we can observe how the atmosphere carries the moisture along the planet, how vapour is invisible, only with satellite images can we follow this phenomenon. When we apply this data to a model with the shape of the Earth, new perspectives are obtained, every global climate is conducted by a single process, the region around the equator receives the highest incidence of solar energy, producing about 65% of all steam, which always travels in the same way feeling towards the poles, conducted by dominant winds and planetary rotation. In the northern hemisphere rotating clockwise, large steam spirals extend for more than 3,000 km, already in the southern hemisphere rotating counterclockwise, the Earth is in search of a balance that will never reach. As these steam- laden winds reach the continental masses of the planet, specific climatic conditions are produced in each location. We can cite as an example the end of July in West India, the hot and humid air pushed upwards by a layer of mountains called *Catis,* gigantic clouds rise, the result of this phenomenon are the *monsoon* rains, trillions of tons of water fall from the sky, transforming the dry region into fertile plains, in China, thanks to these rains, thousands of rice paddies are benefited, bringing food to more than 3.6 billion people, almost half the world population. On the other hand.

On the side of the globe, the winds need to cross the immense Andes to reach the central part of Chile. The altitude eliminates the humidity from the air originating one of the driest regions in the world, the Atacama Desert, with a point that has never been registered the occurrence of rains. Steam is one of the main maintenance forces in the world, but it is only one of a much more complex system.

The icy temperatures at the poles and hot in the equator have a variation of more than 72° C, thanks to these variations, all the air and all the water around the planet are conducted, creating invisible and unexpected mechanisms for the maintenance of life on Earth.

To understand the next component and analyze it from another extraordinary perspective, we need to go to the South of the planet.

Near the Antarctic region, where the plaga suffers the influence of an immense whirlwind with continental proportions, one of the most relevant examples occurs in the waters at 60° South, are the gale of the 60° latitude, the most agitated and aggressive seas on Earth, where persistent winds and storms flog the Antarctic Ocean with an incessant fury, and stirring up more than 130 million tons of water per second, this entire process is driven by the movement of heat that travels from the equator to the poles.

Antarctic Continent (NASA Image, Satellite Aqua)

Chapter 2 - Antarctica

Before we continue our study, it is essential that we know the differences between the Arctic continent and the Antarctic continent, let us analyze the image below...

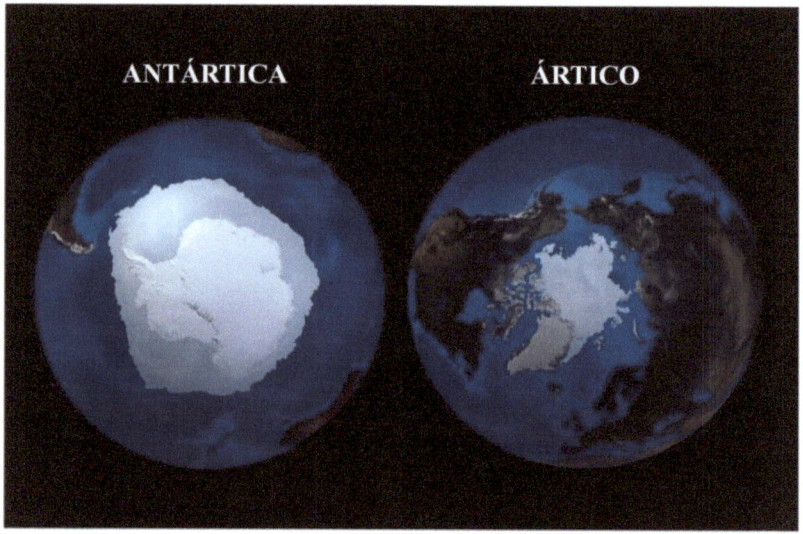

Source: Goddard SpaceFlightCenter of NASA

Some peculiarities between the two continents are; the *Arctic* has no land mass, it is a continental mass of ice floating over the ocean, it is integrated with eight islands around it, they are;

Greenland, Ellesmere Island, Vitoria Island, Bank Island, Wrangel Island, Sévernaya Zemlyá Island, Francisco José Land, Spitsbergen. In this region we can find the majestic Icebergs and the famous Glaciers. The population living in the northern continent is very varied, formed by people who settled in the Bering Strait and Greenland. There are approximately 135 thousand people living in this region. The most characteristic Fauna of the Arctic are the polar bears, which come year after year, reducing their contingent due to climate changes and lack of food. The climate in the Arctic has great variations throughout the year. Located at the extreme north of the planet and due to the inclination of the Earth's axis, some points remain plunged into obscurity during the winter. Even in summer, the sunlight reaching the region is low, so the solar energy is little having much of it reflected back into space by the color of the ice. Throughout the year, o The Arctic radiates more heat than it receives, and most of its heat comes from the tropics through atmospheric and maritime circulation. Scandinavia is the warmest Arctic region due to the influence of the Gulf Stream.

Winters are long and cold, and summers are short and cool, but there are important regional differences . The atmospheric humidity is generally low and precipitation is scarce, some areas receive less than 50 millimeters of rain per year. In summer the rain does not tend to evaporate quickly due to low temperatures and frozen soil (permafrost), prevents its absorption, creating large areas of swamp. The thawing of winter snow also contributes to this, and floods are frequent in large proportions. The accumulation of snow in winter is very variable and depends mainly on geography, atmospheric humidity and wind intensity.

The Arctic has been affected by climate change, leading to the retraction of the frozen cap over the Arctic Ocean, and the release of melted *permafrost*. In September 2007, *ENVISAT,* the largest melt in the Arctic Ocean, was recorded by *ESA* (European Space Agency) satellite. For some years now, there has been a galloping melt in the Arctic area, about half of Greenland's ice sheet melts in summer in its layer superficial, but in the year 2012, 97% of the area of the mantle showed degrees of melting that reached higher and colder parts, a phenomenon that increases the risks of an environmental catastrophe and increasing the speed of displacement of glaciers towards the sea, having as an immediate consequence, Arctic.

In the *Antarctic* continent its severe seas hold a surprising secret that affects the whole world. With an extension of 14.000.000 km², in winter remaining in a total darkness about six months a year, its temperatures reach an average of (-93,2 °C) negative, in summer, its averages are -10 °C in the coastal region and in the interior is -40°C, a completely hostile place, where it has its majority uninhabited and unexplored. Many authors classify this place with "Polar Desert" due to its very low rainfall, winds of 100km/h are common in Antarctica and lasting for weeks, with records of windstorms above 320km/h. Its fauna is limited to penguins (*Spheniscidae*) scientific name, its flora has great difficulty for the development of vegetables due to strong winds, low soil thickness and limited amount of sunlight during winter. For this reason, the variety of species on the surface is limited to "inferior" plants, such as mosses and hepatics. In addition, there is an autotrophic community, formed by protists. The continental flora consists of lichens, bryophytes, algae and fungi. Growth and reproduction usually occur in summer. There are about 230 species of lichens and about 54 species of bryophytes. On the continent there are 712 species of algae, most of which form the phytoplankton. Diatoms and algae from Snow, microscopic algae that grow on snow and ice giving them color, are abundant in coastal regions during the summer.

Currently there are scientists from several countries studying the continent, for a greater understanding of the global importance of this local ice. With these collected data and with the help of satellites, they have come to the conclusion that a number of particularities make the region the coldest on the planet, and with these results, we can conclude that this continent maintains all forms of life on Earth, including the abundant forests that are thousands of miles away. With the joining of fragments of data obtained by 17 different satellites, a powerful climatic system was observed that surrounds this entire continent. A huge whirlwind driven by the Earth's rotation, and as the air hot and humid migrates to the south of the planet, potentiates and forms a gigantic invisible system called *Polar Jet*. The relentless wind drives the sea water down, and the Antarctic Ocean passes the only parallel in the world that has no land, and as a result, an immense circular current spins incessantly, this is the strongest ocean current on the planet, creating the famous 60° latitude gale that intensifies with the combination of water vapor, winds and the shape of the Earth. The *Polar Jet* is so powerful that it isolates Antarctica from the rest of the world, preventing heat and humidity from reaching its interior, giving rise to the driest and windiest region of the globe. Here the blizzards are not caused by the precipitations that come from the sky but by the winds that lift the ice from the ground, this dense and freezing air is a result of the Polar Jets that is able to cool the whole continent. In winter, the conditions even more severe, trigger an essential process for life that occurs under the ice. This process, far away and invisible to the eyes of any human being, something extraordinary happens, having an effect all over the world, every winter in Antarctica, are formed 25 thousand tons of banquisas reaching an area larger than Australia. With the data placed in a model, we can analyze the loss and gain of continental mass in a period of two years, this is the main seasonal change on Earth, producing profound effects on life around the planet. This whole process occurs, thanks to the physical characteristics of salt water. In a remote area of the coast called the *Weddell Sea,* a series of pollinia are formed, extensive areas of sea water surrounded by ice, with catabatic winds cooling the sea water to temperatures below zero. When the temperature in the upper ocean layer reaches -1.5°C, a dangerous boundary is crusade. Now all this command is assumed by another peculiarity of the salt water, on the surface the sea starts to freeze, microscopes crystals begin to grow and intertwine, to freeze totally, the water needs to get rid of salt, the water that remains liquid becomes moresalty, forming a brine that drips through the long tubes created by the newly formed ice. This brine is denser than ordinary salt water and occupies the deepest spaces of the ocean, this denser water carries with it the oxygen present in the surface air leading to the depths.

The formation of ice becomes faster and more intense and in a short time large blocks of flat ice begin to float on the surface forming a rigid mass, in just seven days the microscope process can already be analyzed by satellites, with their sensors and submarines present for this study, in the revealing an extraordinary transformation bringing a consequence although it can never be studied before. Every second, 1.5 million cubic meters of dense and salty water descends to the bottom of the sea, in an uncontrollable vertical current, this water when it reaches the bottom of the sea, spreads for hundreds of kilometers, forming a waterfall on the continental platform, an immense underwater waterfall emerges that has never been seen by a human being, with torrents equivalent to 500 times the Niagara Falls. The cold, dense and oxygenated brine runs slowly and silently to the oceanic abyss, this water, will take about a thousand years to return to the surface again.

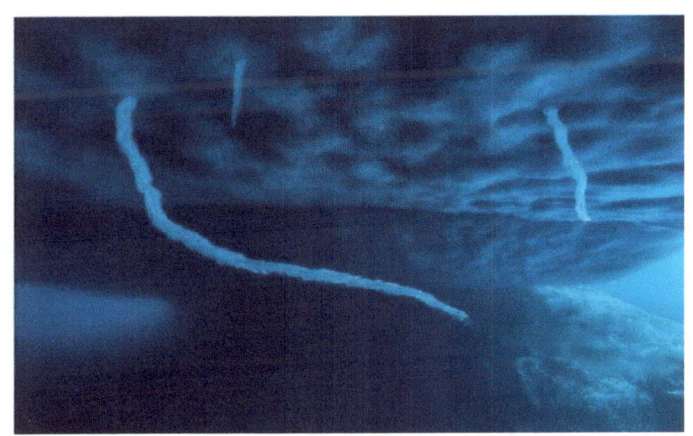

With a combination of data within a mathematical model, shows us the flow of this water back to the equator, migrating to the north of the planet making the oceans colder and more agitated, this system regulates the average temperature by 0.5 C. This stability allows life to flourish by protecting it from drastic changes in climate of the planet. When the deeper waters finally return to the surface, the hotter and faster currents come together, becoming more dynamic. Through analysis, the ocean shows itself as a single mass in an incessant whirlwind, the temperatures of these surface currents vary with the energy received by the sun and with these variations are determined the amounts of steam that will be released into the air and cause seasonal changes on both continents and oceans. In autumn, as the Gulf Streams become colder, the Edges trees change their color to a redder shade and begin to lose their leaves, six months later, on the other side of the world, the *Kuroshio Stream* begins to become warmer allowing the cherry trees to bloom all over Japan. Similar processes occur around the globe, determining cycles seasonal of almost all forms of life on Earth.

Through computational analysis, we can conclude that the ocean and the atmosphere are intimately connected, a continuous system united by more than twelve trillion tons of water that float all over the air uninterruptedly.

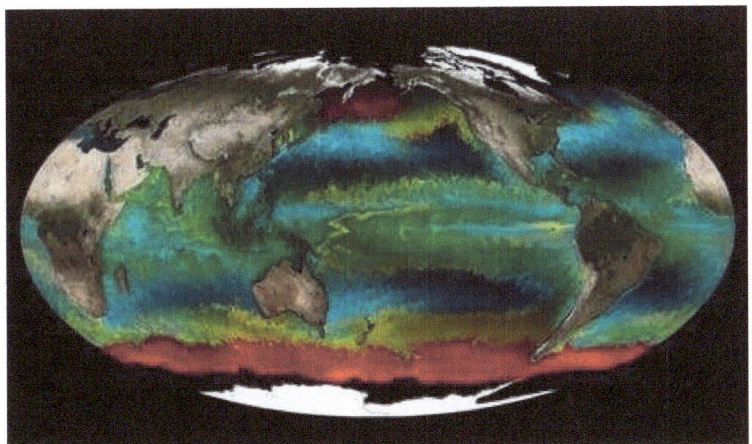

In green, representation of water vapor around the planet.

Each storm, each small drop of water, is part of this complex gear that drives all the activities that form the Our world, however, still has much more in this planetary mechanism than is imagined. Facing one of the most violent systems on Earth, the icy brine of Antarctica is undergoing yet another transformation. At the meeting point between fire and water, something fascinating happens, a process that sustains almost all life in the world.

To the west of Peru, the sea is ravished by a feeding *frenzy*... Plankton serve as a feast for millions of sardines and anchovies, each dwarf, thousands of predatory fish and seabirds migrate to the region to feed on these shoals, is one of the largest volumes of marine life on the planet, also becoming an extremely attractive area for fishing, but this is much more than a rich place for fishing activity, is mainly, one of the best examples of how two of the systems of the Earth are able to interact a prolific of life.

The first of this system is the water cycle, the other is found in the hot and bubbling interior of the planet. From here originate almost all the other substances necessary for the constitution of life, the world is not a solid sphere formed only of rocks, but a burning circle of molten liquid with a cold crust outside. The surface of the Earth is like a coating of a drop of rain, unstable in nature.

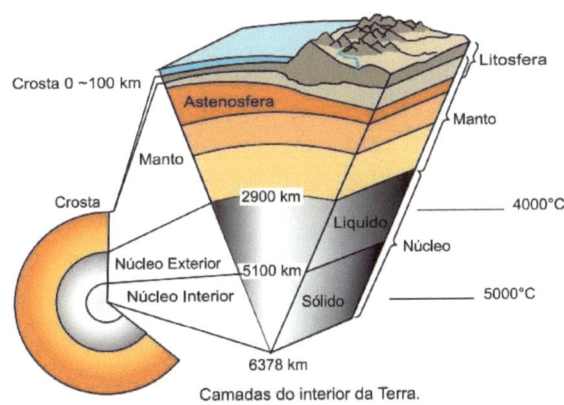

Camadas do interior da Terra.

March 2011, an earthquake with magnitude nine on the *Richter* scale hits the city of Sendai, capital of Miyagi Prefecture in

Japan, the earthquake was so strong that it threw parts of the country 2.5 meters towards North America. Simultaneously, a volcano erupts, a huge pyroclastic ash cloud rises towards the stratosphere. These violent events, are just local disorders, caused by the ancient and slow currents of molten rocks that circulate all the time in the interior of the planet, supplied by the weakening of radiation in the center of the Earth. The substance that leaks through the crust, provides basic elements necessary for life, two systems, one of fire and the other of water, that interact in several places and the most important meeting of all this takes place at the bottom of the sea.

Chapter 3- Planktons and Phytoplanktons

In the depths of the Atlantic Ocean, 2,500 meters from the surface, a chain of submarine volcanoes is hidden, here everything is invaded by lava and overheated gases, the end of a journey of 25 million years from the distant center of the Earth. Here, acidic and toxic whose pressure is hundreds of times higher than on the surface, occurs the basic chemistry of life, gases that would normally evaporate, react vigorously with the dense and oxygen-rich waters from the Antarctic, sea the hot minerals that have traveled the interior of the planet for millions of years dissolve in the sea water. At this time, there is a reaction with oxygen, becoming rich in nutrients.

The oceanic waters now filled with minerals from the interior of the Earth emerge from the hydrothermal vents, living beings struggle to use these waters, bacteria are the first to colonize these vents. They are very fertile conditions for the development of these tiny organisms. Then, more complex creatures begin to feed on these microorganisms and they in turn feed on themselves, the abundance is such that an enormous amount is left over from this process, so the ocean currents take charge of transporting the surplus around the world until they finally reach the surface of the sea. Other currents erode the continental masses of the planet and extract minerals directly from the rocks.

Returning to the famous fishing grounds of the Peruvian region, deep ocean currents are driven upwards as approach the South American continental masses, bringing with them an abundance of nutrients. Phytoplankton, microscopic plant organisms that voraciously consume sunlight and rich water, carbon dioxide is dissolved in the air, providing these single-celled creatures with everything they need to grow and reproduce. At this time, they multiply exponentially, reaching billions of units that can be captured by satellite sensors.

In just 24 hours, 500 square kilometers of blue ocean turns green, phytoplankton growth triggers one of the greatest food frenzy on the planet. The similar emergence of nutrients around the world provides the efflorescence of more planktons, which can be seen through the highest technology, create huge green strips on the globe reaching up to a fifth of the oceans.

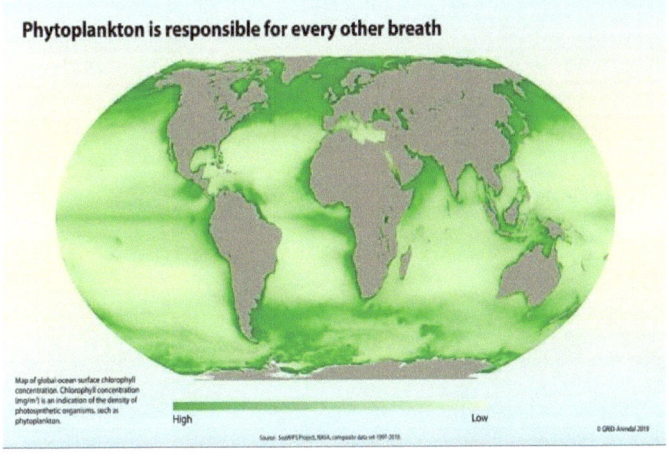

Plankton is the basis of the entire food chain, capable of transporting minerals from Earth directly to all sea creatures, these minerals that once circulated throughout the interior of the planet for millions of years, are now essential instruments for this ocean balance. In the next 24 hours, the planktons that did not serve as food, submerge again, taking with them to the depths the carbon and minerals ingested during the journey, remaining in the ocean floor for thousands of years, forming a thick layer of tiny carcasses up to a kilometer thick, in the future, most of these will emerge again in a second stage, providing the chemical substances necessary for continuity of life on Earth.

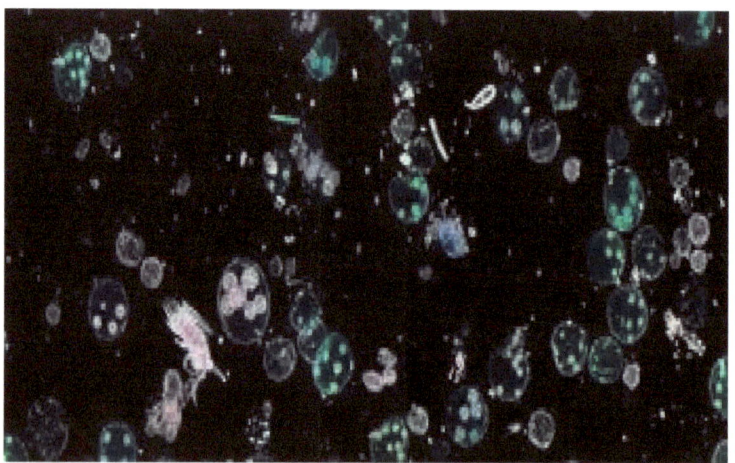

This process plays a fundamental role in the formation of the food we consume and the air we breathe, besides, it supplies the richest ecosystem on the surface of our planet, the Amazon Forest. To understand how this whole process works, we will have to go to one of the driest and most dusty places on Earth, the violent Sahara desert.

The Earth's systems operate in different ways, some as the climate is more dynamic, others as the Earth's core takes a few millennia to complete a single cycle. With the most advanced technology, we can understand how the slow and the fast walk side by side generating extraordinary results.

The Sahara Desert on the African continent is a dry territory, but one day it was green and exuberant, even today it plays a fundamental role in the life cycle of the Earth. In the month of May, the height of the driest season, travelers travel on their camels through one of the most dangerous regions of the Sahara, the *Bodélé Depression,* an ancient sea that dried up five thousand years ago. The soil called *Diatomite* is obtained from very old waste plankton, rich in iron and phosphorus compounds two essential elements for all living organisms. The most curious fact is that these same grains of sand will in just six days revive a tropical forest eight thousand kilometers away. To begin this rebirth process, it is necessary that only one *diatomite* flake is suspended in the air. The flake is fractured into an extremely fine powder and carried by the winds, quickly the air is filled with more and more microscopic flakes, through the data provided by the *MeteoSat* satellite, reveal a daily movement of dust, appearing a gigantic cloud that emerges directly from the desert. The dust rises every day with impressive precision at exactly noon, which began as a microscopic process in a short time became a great sandstorm. A hundred stories high and hundreds of kilometers wide, the cloud of ancient plankton now blows across Africa, on the west coast the dust is driven up by the prevailing winds giving rise to an epic journey across the Atlantic Ocean, satellites reveal to us that fifty-four thousand tons of dust are transported each day traveling eight thousand kilometers to its final destination, Amazonia. It is here, above the humid tropical forest, that the planktons are reborn in a spectacular way, the minerals present in the dust are dissolved in droplets of water, being conducted by rain to the forest core.

During the rainy season in the region, the incessant precipitation spreads over the jungle a total of forty million tons of African dust, what were once plankton now settle on the ground and the roots of trees revitalizing the forest, the process of fertilization of the Amazon by the Saharan dust remained unknown to mankind until the advent of satellite Earth, With extremely sensitive instruments capable not only of observing the migration of dust from Africa to the Amazon but also of measuring the *forest canopy* through space, it is also possible to make a study with the end of the rainy season in the region and follow the return of the sun, for the first time after six months, the sun shines directly on the forest. The result is an explosion of growth, for each leaf there are three more emerged in a period of ten days, a green wave crosses the continent, the migration of dust from the *Bodéle Depression* to the Amazon is just one of thousands of processes Similar to the distribution of minerals essential to living ecosystems around the world, deserts, mountains and ancient sediments, each element has its own composition penetrating the vital chain of the most varied forms. Each portion of the sole around the planet depends on these processes, the great plains of North America, perfect for the production of corn and wheat are formed from glacial deposits, the delta of the river Ganges in Bangladesh is rich in iron that erodes from the Himalayas being one of the fundamental ingredients for the cultivation of rice, other minerals are transported to the entire planet by air, water and ice, as a consequence of this process, plants become capable of radically reconfiguring our world.

Plants are not only a product of the Earth, they configure a powerful force, capable of transforming the planet to millions of people.

years, they are responsible for the changes in the atmosphere and definition of human beings, shaping many aspects of our bodies and minds.

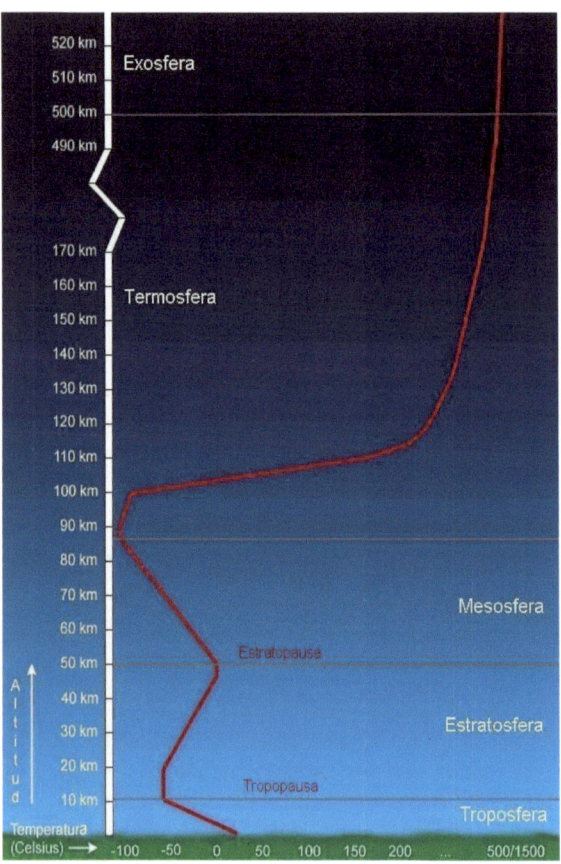

Chapter 4- The Amazon Forest

Another extraordinary process of the planet seen through the satellites, from the analyses made by computers, showing a daily movement of invisible particles of oxygen and carbon dioxide in the air, however, these essential substances for life are not the fruit of a geological process but of trillions of tiny breaths. To understand this system, it will be necessary to return to the Amazon, this humid tropical forest existing at about fifty-five million years of age, is one of the most ancient living ecosystems on Earth, its biodiversity is so unique that it shelters more than half of the living forms of the planet. With a range of six and a half million square kilometers of pure green. Just like Antarctica and the Sahara Desert, this ancient ecosystem plays a key role in the development of the planet. an essential role for the rhythm of life of the whole planet. Here the process begins in the small holes present in the lower parts of the trillions of leaves existing in the forest.

During the day, the leaves absorb the carbon dioxide present in the air, converting it into sugar and releasing the volatile gas we call oxygen.

Evapotranspiration Process

Throughout its life, a single tree is capable of releasing millions of cubic meters of this precious gas, the Amazon processes daily, a fifth of all the oxygen in the world.

For decades it was considered the lung of the world, now, with all the technology of computers and satellites, it is beginning to become clear that nothing of the terrestrial planetary systems is simple. From the analysis of the *Earth* satellite, it was possible to prove that most of the oxygen produced during the day is reabsorbed by the forest itself at night, it takes another step for the excess oxygen to be released.

Every 24 hours, two million tons of sediment are carried from the forest into the vast Amazon River, these sediments, travel for six thousand kilometers eastward reaching the Delta of the Amazon, here the planktons present in the water absorb the sediments, with more sunlight and more carbon dioxide present in the air, the plankton population explodes again. The amount of oxygen released by the planktons is of a gigantic volume, which can be observed from space by our satellites. Half of all the oxygen present in the atmosphere comes from the planktons, these little creatures are the true lungs of the Earth.

Planktons keep the atmosphere in perfect balance and this process enables the next link in the vital chain.

An atmosphere rich in volatile oxygen allows for more dynamic and complex creatures, capable of moving quickly using tails, wings, arms and legs. In reality, the balance of the gases in the air not only defines the size of our bodies, but also determines almost everything we are. However, oxygen also has a negative side, its extreme volatility is capable of provoking violent and uncontrollable reactions, and the most relentless of these is fire, this small detail only shows us a small piece of the complex system that is planet Earth.

Chapter 5- The Fire

October 2013, an immense fire punishes Canada, more precisely in the Yukon territory, an area with a peculiar geography, a mountainous, wild and sparsely populated region, the Kluane National Park and Reserve houses Mount Logan, the highest peak in the country, as well as glaciers, trails and the Alsek River. In less than a week, flames devastate twenty-five thousand kilometers of forest, simultaneously in Siberia, another fire destroys four thousand hectares of forest. All this is a small sample of the unique power of fire around the world.

Every day, the Earth is devastated by huge fires, analyzed by our models as big red spots. Fire is another of the most extraordinary systems on Earth and plays an essential role in the life cycle of the planet.

Boreal Forest, northern Canada is possible fleece in action, this abundant forest of Spruce has a very special relationship with fire, here the extreme cold kills and numbs most of the trees, trapped in these trunks, are the components necessary for the emergence of new forms of life, however, under these conditions, this process would take hundreds of years, but in the presence of fire, it could trigger it in a matter of hours.

Fir Forest (Canada)

Most natural fires start from random electrical discharges from the sky, the Spruces, are a perfect fuel for fire, their combustion is easy and fast that a small spark is able to make them burst into flames. In this way, the volatile oxygen undoes their lethal blow, the hot oxygen binds to the carbon atoms present in the wood of the trees generating more heat making the bonding of oxygen with new carbon atoms faster and generating much more heat causing the flames to intensify. As the fire devours everything around it, the solar energy that was stored inside the plants is released, this is the dynamic of the fire.

To observe a burning flame is to witness the power of the sun as it frees itself from the life that imprisoned it for a long time, in a matter of hours, what begins with a small spark takes hundreds of hectares of forest into flames. The organic matter stored by these trees for hundreds of years quickly turns to ashes, these flames eliminate dead and sick organisms from the forest by recycling them and restoring their minerals to the ground.

As we observe the fire of this prism, it is nothing more than the part of a rebirth and regeneration. Fire has existed since the evolution of plants, at the same time that they began to produce oxygen, they made possible the production of the substances necessary for combustion, besides making possible the existence of fire, many plants also depend on it, the Firs, for example, evolved in such a way as to release their seeds in the middle of the ashes that accumulate in the soil after a fire.

Through the satellites in the Earth's orbit, it is possible to visualize the effects of fires around the world, after each one of them, what follows is the tendency of a new growth of life, preserving the health and promoting the regeneration of diverse ecosystems of the world, avoiding in a unique way their stagnation.

Satellites reveal to us how fire, climate, water and ice are associated for the maintenance of the life cycle, everything is interconnected in a millennial and complete system, but this is only the beginning of the discoveries made through the new technologies. With this, we are able to analyze, explore and identify any external reaction that shows us with conviction, that no element can exert greater influence on the planet than the sun.

Chapter 6 - The Sun

During the 24 hours it takes the Earth to perform its rotation movement, it reacts to the extraordinary forces of the sun, each day, 170 million *gigawatts* (GW), which corresponds to seven thousand times the energy consumed by humanity, are poured into the surface of the planet, triggering a ceaseless wave of activity.

At dawn plants and planktons begin the process of photosynthesis, using sunlight they produce sugars and starches that are the basis of the food chain and the main source of energy for almost all living beings.

Sunlight controls the winds and weather around the globe at night, when the air cools down many rains are triggered. We too are part of this circadian cycle and we respond to the flow of energy that comes daily from the sun. To produce vitamins in the skin, the cells of our body need sunlight, even the routes of flights reveal a close relationship with the sun, during the mornings, the aircrafts travel westward to extend the day and in night flights, travel eastward, for the purpose of shortening the night.

The irony, however, is that the threat to this harmonious system comes from the same place that allowed its existence, the energy emitted by the sun.

Based on the analyses of the *SDO* satellite, an infrared record of the radiation released by our star, are thoroughly studied. Charged particles, proton fractions, electrons and neutrons are discarded all the time along with huge pulses of electromagnetic radiation.

Sporadically, the sun discards coronal mass ejection, with a supercomputer, it was possible to follow the images of an immense plasma cloud millions of kilometers long towards the Earth.

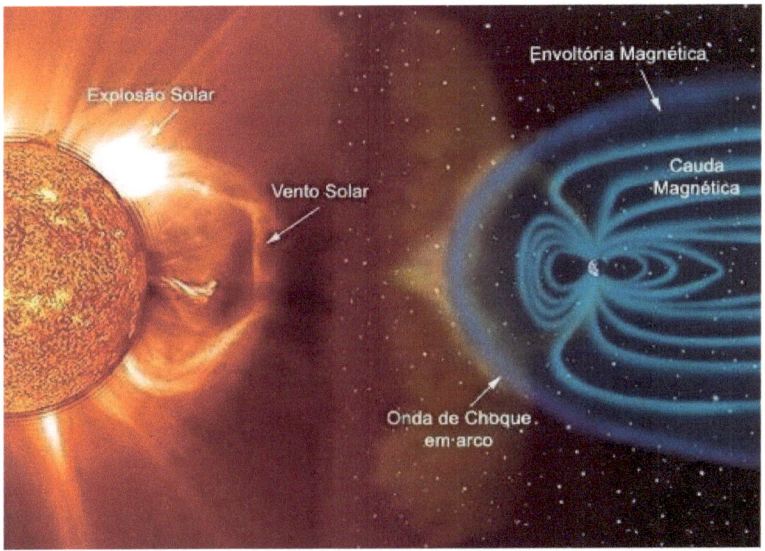

If for an instant, these solar particles were able to reach the surface of the Earth, they would produce fatal mutations in the *DNA* (deoxyribonucleic acid) of all living creatures, causing serious problems in our planet. Fortunately the planet can defend itself.

Our planet is surrounded by an invisible force field called the *Magnetosphere,* with images from five magnetically synchronized satellites, this technological network called *Themis.* A space mission that originally would be a constellation of five satellites identified as:

THEMIS **A**, THEMIS **B**, **THEMIS C**, **THEMIS D** and **THEMIS E**, would study the launching of energy from the Earth's magnetosphere known as sub-tempests, celestial phenomena that intensify the occurrence of auroras near the north and south poles.

Currently, three of the satellites remain in orbit of the Earth, two of them have been diverted to the vicinity the Lunar orbit. Launched in 17 February 2007 from the aerospace launch base at Cape Canaveral, United States, aboard a Delta II rocket. Each satellite carries identical instruments, including a fluxgate magnetometer (FGM), an electrostatic analyzer (ESA), a solid-state telescope (SST), a search-coil magnetometer

SCM) and an electric field instrument (EFI). Each has a mass of 126 kg, including 49 kg of fuel.

They revealed to us our force field constantly bombarded by the sun, the shape of the field is shaped only by the strong attacks of radiation, a nebular lagoon 320 kilometers in diameter, wave after wave, the solar particles reach the magnetosphere, most of them are deflected, but when the field is hit by a coronal mass ejection, the charged particles manage to break its layer more external, in sequence, once they cross the shield, they are free for their advance towards the planet. The magnetic field guides the particles towards the poles, giving rise to one of nature's most impressive spectacles, the northern lights and the southern lights or more popularly known as Auroras Boreais and Auroras Austrais. In the image below it is possible to analyze the second layer of defense of the Earth.

Gigantic strips of plasma form a downward current, surrounding the poles of the planet, as they quickly reach the upper layer of the atmosphere, they agitate the air molecules causing them to begin to shine, oxygen radiates the colors red and green, and nitrogen radiates the color blue. An energy capable of modifying all life on Earth is dissipated by the upper layer of the atmosphere, thus the planet has been able to protect itself for millions of years against mortal radiation from the sun. But even with this extraordinary apparatus, it is only one part of how the atmosphere is able to protect life on Earth.

Images of the Earth's Magnetosphere

Even more powerful systems exist far below, without which life would be impossible.

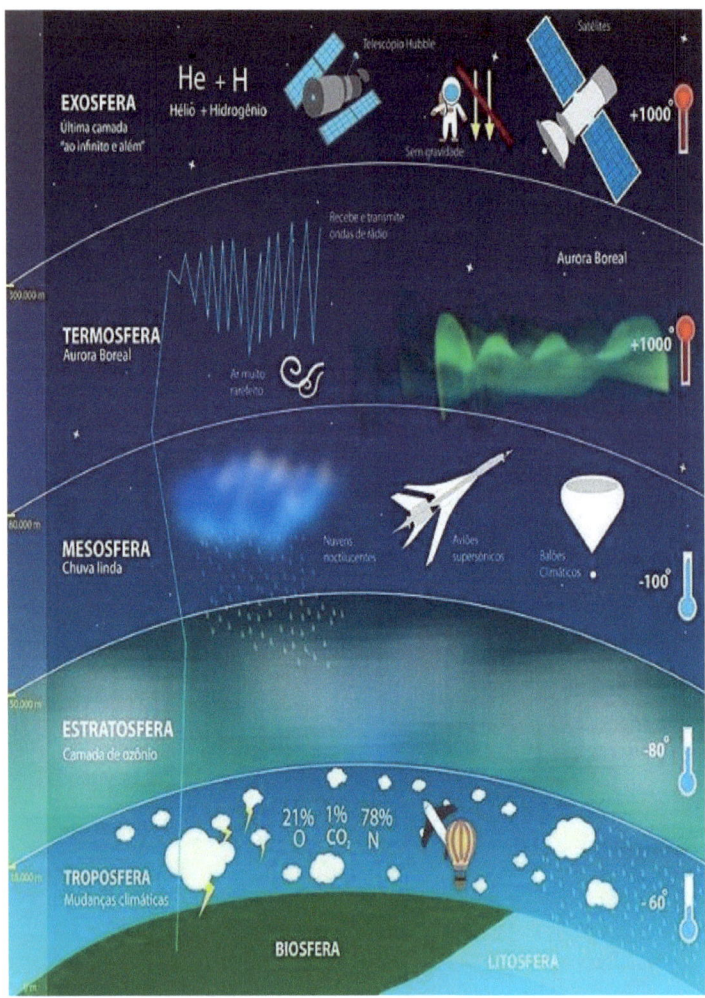

Chapter 7- The Earth Atmosphere

The Earth's atmosphere is a very delicate resource, a thin blue shell capable of encapsulating our entire world. This thin coating of oxygen and nitrogen, is subjected to intense bombardments of sunlight and heat, forces that in cases of uncontrol, are capable of destroying the entire atmosphere.

During the night, these satellites investigate the murmuring of the Earth by means of lightning. With the support of astronauts from the International Space Station (*ISS*), they provide impressive data, a frequent intensity of electrical storms. Why does the planet need and produce these phenomena?

With the use of the highest technology, this answer becomes clear; the Earth's atmosphere is in search of balance. Each day, the combined force of steam and sunlight creates forty thousand clouds, charged with an immense amount of electrical energy. Every thirty minutes, a medium-sized cloud is capable of generating 100 (MW) megawatts, enough energy to supply the city of Campinas for one minute. To balance itself, the cloud discharges negative energy to the ground in the form of lightning, simultaneously releasing a positive charge.

To the top towards the sky, from each cloud, an immense column of charges emerges, this invisible force moves at almost the speed of light towards the outer layer of the atmosphere, the Ionosphere.

This layer is formed by a thin veil formed basically of (H) hydrogen and (He) helium, with the data provided by satellites it is possible to see the interaction of electrical charges with this extremely rarefied field. The Ionosphere acts as an electrical conductor, distributing the charge throughout the planet.

Now we know that life would be impossible without this global electric circuit.

All this is due to an extraordinary chemical reaction that occurs inside the clouds charged with the appearance of lightning. The electrical charge inside the cloud is growing extremely strong that the air is decomposed into ions, consequently a tiny path is formed where an electric current passes. Within thousandths of a second, a ray is fired, its thickness is similar to that of a human thumb, but its temperature is five times higher than the surface of the sun. As it crosses the air, this burning ray of energy destroys the molecules of *(N)* nitrogen, the *(O)* oxygen binds to *(N)* nitrogen originating a substance called *(NO3)* Nitrate.

Daily about fourteen thousand tons of *(NO3)* nitrate are transported around the world, with the rains this substance spreads on the ground being an essential element for almost all forms of life on Earth, from the photosynthesis of plants to the respiration of more complex organisms.

Nitrate (NO_3) has been driving the most important chemical reactions for living beings for millions of years. With data arriving daily, we can conclude an intricate mechanism that configures and reconfigures life at every moment and driving the heartbeats of every human being around the planet. What is even more lacking is a part of this complex system, which is the profound and undeniable consequence of a single animal species, the human race.

Chapter 8 - Human Beings

From all these technologies, it has been revealed to us a hidden and complex system that intertwines on all levels, extremely slow processes connect to others that occur within milliseconds, endless cycles of death, decomposition, regeneration and rebirth fill the world.

From the relentless power of solar energy and water, from the electromagnetic forces that operate around us, each interaction reveals to us a harmony and a precise balance. Humanity is the latest natural phenomenon, we are the direct consequence of a system that has been able to create and maintain life for 3.5 billion years. We have developed intelligence and this fact has allowed us to bring contributions to the oldest processes existing on Earth, humanity has transformed o planet by exploring the same complex system that originated it.

Our ability to control ecosystems has allowed our civilizations to grow rapidly and to become the dominant species. Today it is possible to see the influence of humanity, not only and, 82% of terrestrial territories, but also around space, with trips to the moon and with the International Space Station (*ISS*), now we finally begin to understand how our world works and what place we occupy within it.

This is the crucial moment in Earth's history, by observing the planet through the highest technology, it is possible to see that we have become a global force, we already manufacture more (*NO3*) nitrate than lightning, we release more sulfur into the air than all volcanoes in the world, we emit more dioxide of carbon than the entire Amazon, our cities produce dust, leverage electrical storms and affect rainfall systems.

We have the power to impact on large parts of the Earth's cycles, through analysis, the influence of humanity can be considered a natural process.

The gases released by airplanes, cars, power plants, etc... are effects caused by an animal that the Earth itself has produced.

However, there is a fundamental difference, unlike volcanism, the movements of ocean currents or the oxygen released by forests or planktons, we possess the gift of free will, the technologies besides allowing us the impacts that we cause in the world, they help us to make conscious decisions about consumption continuous of our planet's resources. Our new technological eyes are teaching us to maintain the balance capable of sustaining the natural world.

Bibliographical References

Brazilian Space Agency autarchy of the Ministry of Science, Technology and Innovation

Antarctic Glaciology Program. The National Science Foundation. Consulted on August 19, 2009. Copy filed on October 25, 2019.

ESA European Space Agency

ESA Portal - Satellites witness lowest Arctic ice coverage in history" (em inglês). European Space Agency. 14 de setembro de 2007. Consultado em 26 de julho de 2019

Evidence of Ancient Martian Life in Meteorite ALH84001?" (in English). National Aeronautics and Space Administration. Consulted on August 26, 2009. Filed from the original on August 25, 2019.

Glomsrød, Solveig et alii. "Arctic economies within the Arctic nations". In: Glomsrød, Solveig; Duhaime, Gérard; Aslaksen, Iulie (eds.). *The Economy of the North*. Statistics Norway, 2015, pp. 37-78

JAXA - Japan Aerospace eXploration Agency

NASA National Aeronautics and Space Administration

Neil Glasser of Aberystwyth University. "Antarctic Ice Shelf Collapse Blamed On More Than Climate Change. Consulted on August 20, 2019. Copy filed on December 25, 2015.

NOAA National Oceanic and Atmospheric Administration

Satellites see Unprecedented Greenland Ice Sheet Melt - NASA Jet Propulsion Laboratory". NASA. July 24, 2012. Consulted on July 26,
2019.

Science in Antarctic" (in English). Antarctic Connection. Consulted on February 4, 2020. Filed from the original on February 7, 2006.

The Antarctic Ozone hole, NASA Advanced SupercomputingDivision (NAS)".

Nas.nasa.gov. June 26, 2001. Consulted on February 7, 2020. Copy filed on April 3, 2009.

http://www-loa.univ-lille1.fr/

https://aqua.nasa.gov/

https://aura.gsfc.nasa.gov/index.html

https://cloudsat.atmos.colostate.edu/

https://terra.nasa.gov/

https://www.nasa.gov/mission_pages/sdo/main/index.html

https://www-calipso.larc.nasa.gov/

www.ingramcontent.com/pod-product-compliance
Lightning Source LLC
Chambersburg PA
CBHW040233220526
45473CB00001B/226